MATH masterpieces

Math Skills + Puzzles = Masterpieces

octagon

77.4
8

) 978

rhombus

parallelogram

501
334

gle

Gunter Schymkiw

Published by World Teachers Press®
www.worldteacherspress.com

Order Number 2-5259
ISBN 978-1-58324-236-0

D E F G H 12 11 10 09 08

The Association of Educational Publishers

395 Main Street
Rowley, MA 01969
www.didax.com

FOREWORD

The activities in *Math Masterpieces* integrate mathematics with aspects of visual arts. As well as being provided with opportunities to consolidate knowledge and skills in mathematics, students are introduced to significant works of art and their artists. *Math Masterpieces* aims to enhance children's appreciation of the works of great artists, at the same time allowing them to build mathematical skills.

Also available: *Math Masterpieces, Grades 3-5*

CONTENTS

Each selected artwork is paired with a mathematical concept in a group of three pages.

Page 1 of each group consists of a teacher page. Each teacher page contains the following information.

The **artwork and artist** to which this page relates is given.

Background Information includes information about the artist, his/her style, influences and other artworks.

The **Internet Image Search** encourages teachers and students to more broadly appreciate the works of individual artists.

Student Instructions explain how to complete the activity.

The **mathematical concept** being covered is indicated at the top of the page.

A picture of the **completed artwork** shows students the correct answers to their mathematical activity.

A **"Talking Mathematically"** section is included with each activity. This provides lesson format suggestions, discusses possible mathematical emphases and provides further interesting background information that is both instructional and historical in its nature.

Occasionally, **additional activities** are given which may extend the activity into other learning areas.

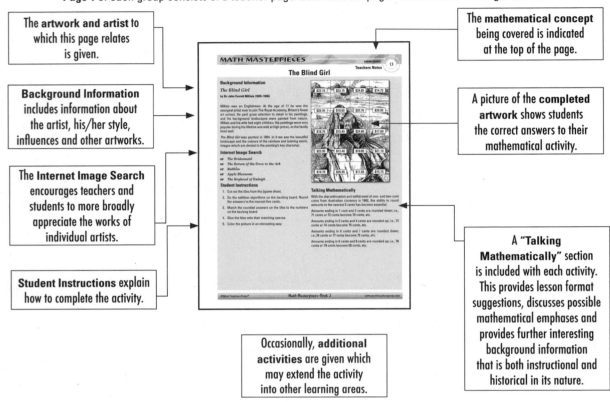

Page 2 of each group consists of the student activity page with the mathematical problems to be solved.

The **mathematical concept** being covered is indicated at the top of the page.

The **artwork** to which this page relates is given.

The **algorithms/problems** for the students to complete are provided.

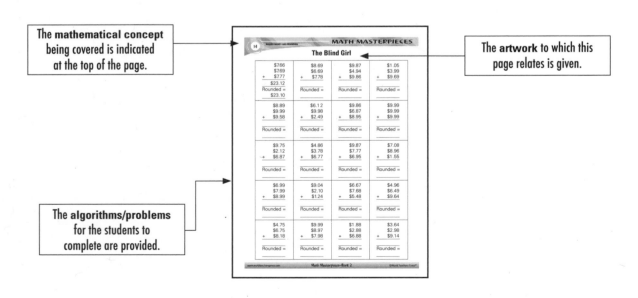

Page 3 of each group provides the answers to the algorithms/problems on "mixed up" sections of the masterpiece.

The **mathematical concept** being covered is indicated at the top of the page.

The **artwork** to which this page relates is given.

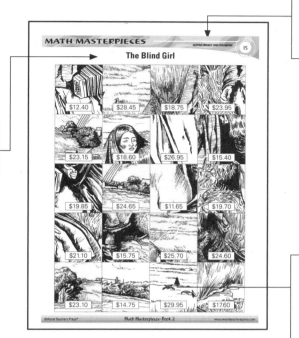

Answers for the algorithms/problems are given on the sections of the masterpiece. These sections will give a completed picture of the masterpiece once glued in the correct position.

NCTM CORRELATION

	Number and Operations	Algebra	Geometry	Measurement
View of Sydney Cove	Place Value			
The Cannon Shot	Roman Numerals			Standard Units
Landscape at Regensburg		Analyze Patterns		
The Cowboy	3-Digit Addition			
The Blind Girl	Estimation			Money
Two Women on the Beach	3-Digit Subtraction			
The Tower of Babel	Place Value			Money
Max Schmitt in a Single	Multiplication			
Study for the Last Supper	Multiplication			
A Winter Scene	Division			
The Peasant Dance	Division			Money
The Avenue	Word Problems			Money, Time
Centurion				Time
The Miraculous Draught	Fractions	Analyze Patterns		
The Feast of Belshazzar	Fractions	Representations		
Bailed Up	Fractions/Decimals	Use Equations		
Bathers at Asnieres	Decimals			Money
The Scapegoat	Decimals			
Self-Portrait	Multi Operations		2-D Shapes	

Teacher Notes

View of Sydney Cove and Fort Macquarie

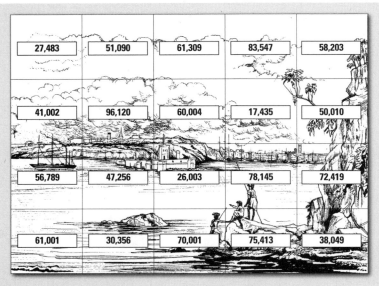

27,483	51,090	61,309	83,547	58,203
41,002	96,120	60,004	17,435	50,010
56,789	47,256	26,003	78,145	72,419
61,001	30,356	70,001	75,413	38,049

Background Information

View of Sydney Cove and Fort Macquarie

by Conrad Martens (1801–1878)

Martens was born in England and came to Australia in 1835. Before this, he had spent two years on the ship *Beagle*, where he worked as an artist for Charles Darwin.

It was as a result of his discoveries and observations on this journey that Darwin proposed his theory of evolution.

Martens' paintings offer a record of early Sydney, and Sydney Harbor was his favorite subject.

Internet Image Search

☞ *The Beagle*

☞ *View of the Heads, Port Jackson*

☞ *Scene on the Nepean River*

☞ *Forest, Cunningham's Gap*

☞ *Terrawambella on Nicol's run*

Student Instructions

1. Cut out the tiles from the jigsaw sheet.

2. Do the addition algorithms on the backing board.

3. Match the amounts on the tiles to the answers on the backing board.

4. Glue the tiles onto their matching spaces.

5. Color the picture in an interesting way.

Talking Mathematically

The Hindu-Arabic system (also called the decimal system) was popularized in Europe by the Italian mathematician Leonardo Fibonacci (1170–1250), who was also known as Leonardo of Pisa. Up until that time Roman numerals were used.

The decimal system uses ten digits (0 1 2 3 4 5 6 7 8 9) to write all numbers. Its basis is a system of place value where each column has a value that is ten times that immediately to its right.

This holds true even with decimal fractions (decimals), where any number to the right of the decimal point has one tenth of its value in the ones column.

Unlike most number systems, the decimal system has a numeral for zero. Zero is used as a place holder in larger numbers.

The method of representing numbers as shown in this activity is called decomposition.

View of Sydney Cove and Fort Macquarie

50,000 8,000 200 + 3	50,000 + 10	70,000 2,000 400 10 + 9	30,000 8,000 40 + 9
80,000 3,000 500 40 + 7	10,000 7,000 400 30 + 5	70,000 8,000 100 40 + 5	70,000 5,000 400 10 + 3
60,000 1,000 300 + 9	60,000 + 4	20,000 6,000 + 3	70,000 + 1
50,000 1,000 + 90	90,000 6,000 100 + 20	40,000 7,000 200 50 + 6	30,000 300 50 + 6
20,000 7,000 400 80 + 3	40,000 1,000 + 2	50,000 6,000 700 80 + 9	60,000 1,000 + 1

MATH MASTERPIECES

View of Sydney Cove and Fort Macquarie

The Cannon Shot

Background Information

The Cannon Shot

by Willem van de Velde the Younger (1633–1707)

Willem van de Velde the Younger lived at a time when Holland was a powerful country in world affairs because of the strength of its navy.

His specialty as an artist was the painting of ships and seascapes. Later in his career his pictures featured battle scenes at sea and ships in stormy settings. His early life was spent in Holland, the country of his birth.

His father was also an artist whose specialty was marine painting. In 1672, he and his father moved to England where they were invited to set up a studio in the Queen's House. They received many commissions for paintings from royalty.

Internet Image Search

☞ *Ships in a Calm Sea*
☞ *Ships off the Coast*
☞ *The Gust*
☞ *Entrance to a Dutch Port*

Student Instructions

1. Cut out the tiles from the jigsaw sheet.

2. Match the numbers on the tiles to their Roman numeral equivalents on the backing board.

3. Glue the tiles onto their matching spaces.

4. Color the picture in an interesting way.

Talking Mathematically

The system of writing numerals used by the Romans involved using seven letters.

These are:

$$1 = I, V = 5, X = 10, L = 50, C = 100,$$
$$D = 500, M = 1000$$

Where a smaller number follows one that is larger, the numbers are added. If a smaller number precedes one that is larger, it is subtracted.

So:

$$VI = 6 (5 + 1) \text{ and } IV = 4 (5 - 1);$$
$$LX = 60 (50 + 10) \text{ and } XL = 40 (50 - 10);$$
$$CX = 110 (100 + 10) \text{ and } XC = 90 (100 - 10);$$
$$MCMXCIX = 1999$$
$$(1,000 + [1,000 - 100] + [100 - 10] + [10 - 1]).$$

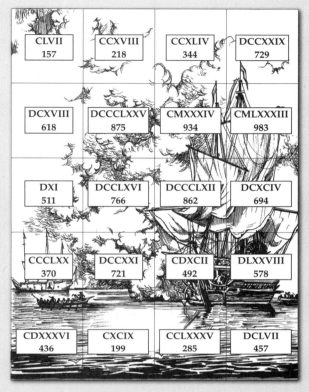

CLVII 157	CCXVIII 218	CCXLIV 344	DCCXXIX 729
DCXVIII 618	DCCCLXXV 875	CMXXXIV 934	CMLXXXIII 983
DXI 511	DCCLXVI 766	DCCCLXII 862	DCXCIV 694
CCCLXX 370	DCCXXI 721	CDXCII 492	DLXXVIII 578
CDXXXVI 436	CXCIX 199	CCLXXXV 285	DCLVII 457

The generally accepted numeral for four is **IV**. Many clock faces using Roman numerals, however, represent four using four strokes (**IIII**). There are a number of explanations for this, but nobody can prove any of them to be the exclusive truth.

One explanation is that the four strokes make the clock face appear balanced and symmetrical (balancing with the four strokes needed for **VIII**).

Another dates back to 1364, when King Charles the Wise of France chided watchmaker Henry De Vick for writing **IV** on a tower clock face rather than **IIII**. Charles cut all argument short by simply declaring, "I am never wrong!"

The most famous clock face of all, that of Big Ben's clock tower in London, uses **IV** for four.

The Roman system of writing numerals does not have a symbol for zero.

Numbers in the thousands beyond 3000 are indicated by placing a stroke above them.

For example, 4000 is written $\overline{IV}$.

MATH MASTERPIECES

The Cannon Shot

CLVII	CCXVIII	CCCXLIV	DCCXXIX
DCXVIII	DCCCLXXV	CMXXXIV	CMLXXXIII
DXI	DCCLXVI	DCCCLXII	DCXCIV
CCCLXX	DCCXXI	CDXCII	DLXXVIII
CDXXXVI	CXCIX	CCLXXXV	CDLVII

Math Masterpieces–Book 2

The Cannon Shot

Landscape at Regensburg

Background Information

Landscape at Regensburg

by Albrecht Altdorfer (c.1480–1538)

Altdorfer lived in Bavaria, Germany. His paintings were unusual for the time in which he lived because most did not contain human figures. Even though there is little human presence in his paintings, the landscapes he painted conveyed an image of foreboding drama. *Landscape at Regensburg* (the artist's hometown) shows a wild image of nature with gathering clouds warning of an impending storm.

Internet Image Search

☞ *Susanna at Her Bath*
☞ *The Stigmatization of St. Francis*
☞ *The Crucifixion*
☞ *Landscape with Satyr Family*
☞ *Alexander's Victory (The Battle at Issus)*

Student Instructions

1. Cut out the tiles from the jigsaw sheet.

2. Match the numbers on the tiles to their Chinese numeral equivalents on the backing board.

3. Glue the tiles onto their matching spaces.

4. Color the picture in an interesting way.

Talking Mathematically

Before the students begin the activity, write these sample Chinese numerals and their Hindu-Arabic equivalents on the board. Note there is no symbol for 0. Challenge the students to explain the system

1	2	3	4	5	6	7	8
一	二	三	四	五	六	七	八
9	10	100	642	738	951	709	560
九	十	百	六百四十二	七百三十八	九百五十一	七百九	五百六十

Landscape at Regensburg

三百九十二	六百五十八	三百四十五	五百八十三
三百六十五	六百八十七	八百六十三	六百二十七
二百八	五百四十九	三百二十七	六百二十三
九百九十	三百七十四	八百三十五	九百七十四
二百四十五	五百九	二百七十六	四百五十二

MATH MASTERPIECES

Landscape at Regensburg

The Cowboy

Background Information

The Cowboy

by Frederic Remington (1861–1909)

Remington was an American painter and sculptor. He is famed for his paintings and bronze sculptures that depicted life in the American West of the late 19th century.

Internet Image Search

☞ *On the Southern Plains*

☞ *Scout*

☞ *His First Lesson*

☞ *Stampede*

☞ *Fall of the Cowboy*

☞ *The Outlier*

☞ *Dash for Timber*

Student Instructions

1. Cut out the tiles from the jigsaw sheet

2. Do the addition algorithms on the backing board.

3. Match the numbers on the tiles to the answers on the backing board.

4. Glue the tiles onto their matching spaces.

5. Color the picture in an interesting way.

Talking Mathematically

Students will need a sound knowledge of trading ones to tens, tens to hundreds, and hundreds to thousands to successfully complete this activity.

MATH MASTERPIECES

The Cowboy

697 284 326 + 112	497 183 674 + 215	978 149 396 + 278	675 184 296 + 378
475 298 378 + 647	778 879 967 + 473	786 365 452 + 623	778 876 543 + 210
907 579 615 + 897	567 678 789 + 891	456 786 856 + 473	784 459 944 + 763
969 748 287 + 693	208 916 317 + 820	971 284 685 + 768	676 149 278 + 385
784 486 693 + 376	685 587 763 + 812	424 168 843 + 279	679 978 411 + 625

The Cowboy

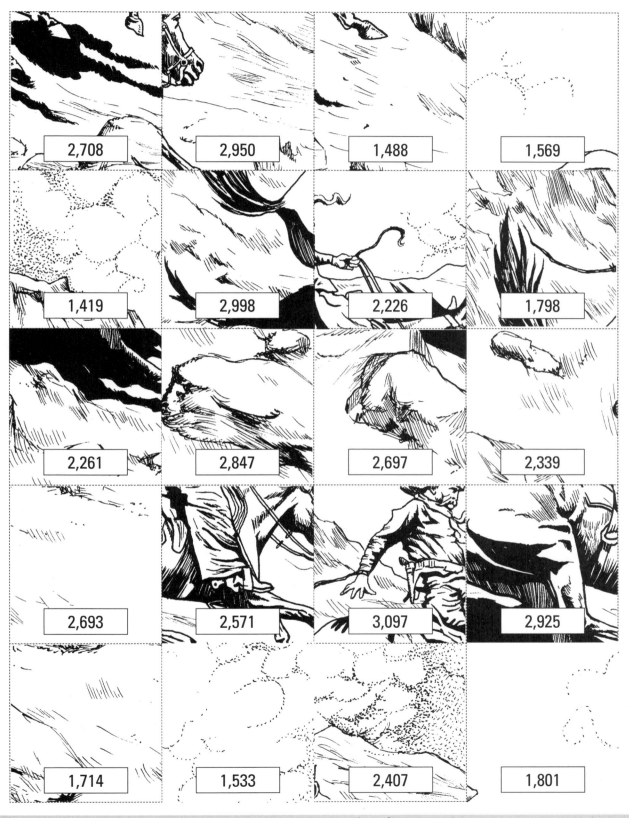

2,708 2,950 1,488 1,569

1,419 2,998 2,226 1,798

2,261 2,847 2,697 2,339

2,693 2,571 3,097 2,925

1,714 1,533 2,407 1,801

Teacher Notes

The Blind Girl

Background Information

The Blind Girl

by Sir John Everett Millais (1829–1896)

Millais was an Englishman. At the age of 11 he was the youngest artist ever to join The Royal Academy, Britain's finest art school. He paid great attention to detail in his paintings, and his background landscapes were painted from nature. Millais and his wife had eight children. His paintings were very popular during his lifetime and sold at high prices, so the family lived well.

The Blind Girl was painted in 1854. In it we see the beautiful landscape and the colors of the rainbow and looming storm, images which are denied to the painting's key character.

Internet Image Search

- ☞ *The Bridesmaid*
- ☞ *The Return of the Dove to the Ark*
- ☞ *Bubbles*
- ☞ *Apple Blossoms*
- ☞ *The Boyhood of Raleigh*

Student Instructions

1. Cut out the tiles from the jigsaw sheet.
2. Do the addition algorithms on the backing board. Round the answers to the nearest five cents.
3. Match the rounded answers on the tiles to the numbers on the backing board.
4. Glue the tiles onto their matching spaces.
5. Color the picture in an interesting way.

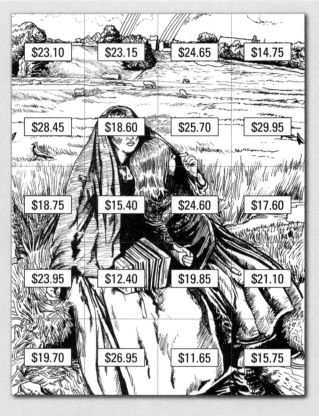

Talking Mathematically

In this activity, students must round to the nearest nickel.

Amounts ending in 1 cent and 2 cents are rounded down; i.e., 71 cents or 72 cents become 70 cents, etc.

Amounts ending in 3 cents and 4 cents are rounded up; i.e., 73 cents or 74 cents become 75 cents, etc.

Amounts ending in 6 cents and 7 cents are rounded down; i.e., 76 cents or 77 cents become 75 cents, etc.

Amounts ending in 8 cents and 9 cents are rounded up; i.e., 78 cents or 79 cents become 80 cents, etc.

The Blind Girl

$7.66 7.69 + 7.77 $23.12 Rounded = $23.10	$8.69 6.69 + 7.78 Rounded =	$9.87 4.94 + 9.86 Rounded =	$1.05 3.99 + 9.69 Rounded =
$8.89 9.99 + 9.58 Rounded =	$6.12 9.98 + 2.49 Rounded =	$9.86 6.87 + 8.95 Rounded =	$9.99 9.99 + 9.99 Rounded =
$9.75 2.12 + 6.87 Rounded =	$4.86 3.78 + 6.77 Rounded =	$9.87 7.77 + 6.95 Rounded =	$7.08 8.96 + 1.55 Rounded =
$6.99 7.99 + 8.99 Rounded =	$9.04 2.10 + 1.24 Rounded =	$6.67 7.68 + 5.48 Rounded =	$4.96 6.49 + 9.64 Rounded =
$4.75 6.75 + 8.18 Rounded =	$9.99 8.97 + 7.98 Rounded =	$1.88 2.88 + 6.88 Rounded =	$3.64 2.98 + 9.14 Rounded =

The Blind Girl

Two Women on the Beach

Background Information

Two Women on the Beach
by Paul Gauguin (1848–1903)

Before devoting his life to painting, Gauguin was a successful stockbroker who had been a patron of the arts. He began exhibiting his own work in 1876 and in 1883 left his job to concentrate on painting.

He was a friend of van Gogh, but the friendship was a stormy one. Gauguin was unable to support his family as a painter. He longed for a simple life and moved to Tahiti in 1891. *Two Women on the Beach* was painted while he was living in Tahiti.

Internet Image Search

☞ *Vision after the Sermon*

☞ *D'ou venons-nous?*

☞ *Washerwomen at Pont-Aven*

☞ *Night Cafe at Arles*

☞ *Siesta*

☞ *The Swineherd*

☞ *Cattle Drinking*

Student Instructions

1. Cut out the tiles from the jigsaw sheet.

2. Do the subtraction algorithms on the backing board.

3. Match the numbers on the tiles to the answers on the backing board.

4. Glue the tiles onto their matching spaces.

5. Color the picture in an interesting way.

Talking Mathematically

Your students will need a good knowledge of regrouping to do these examples.

Using 501 minus 334, follow these steps:

Step 1: 1 minus 4 can't be done so you must borrow or regroup.

Step 2: Look in the tens column. The number there is "0" so you can not borrow from there.

Step 3: Look in the hundreds column. Exchange 5 hundreds for 4 hundreds and 10 tens.

Step 4: You now have 10 tens in the tens column. Regroup these for 9 tens (in the tens column) and 10 ones (in the ones column).

Step 5: There are now 11 ones in the ones column, so the subtraction in this column can now proceed. 11 minus 4 equals 7—write "7" on the answer line under the ones column.

Step 6: After regrouping, the tens column now reads 9 (tens) minus 3 (tens). Do the subtraction and write 6 on the answer line in the tens column.

Step 7: After regrouping, the hundreds column now reads 4 (hundreds) minus 3 (hundreds). Do the subtraction and write 1 under the hundreds column.

Demonstrate to students that after regrouping, 501 becomes 4 hundreds, 9 tens and 11 ones. Add these numbers (400 + 90 + 11). The answer is 501. Regrouping has reorganized the composition of the original 501 (5 hundreds, 0 tens and 1 one).

MATH MASTERPIECES

Two Women on the Beach

900 − 456	911 − 302	439 − 251	821 − 203
845 − 156	624 − 313	951 − 145	941 − 576
611 − 377	963 − 148	713 − 451	803 − 696
749 − 221	911 − 194	902 − 334	858 − 429
501 − 334	679 − 381	863 − 290	902 − 748

Two Women on the Beach

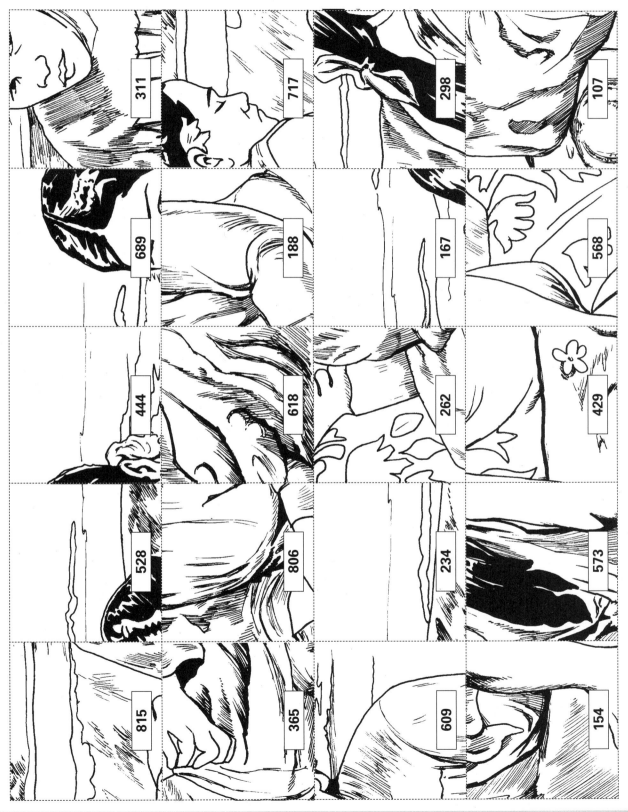

The Tower of Babel

Background Information

The Tower of Babel

by Pieter Bruegel (c.1525–1569)

Pieter Bruegel the Elder is the best known of a family of Dutch artists. His paintings are very detailed and busy.

Children's Games is a scene showing more than 200 children at play. *The Tower of Babel* depicts a scene from the Bible story in which the building of a tower built to reach heaven was frustrated by God who confused the builders' languages so they could not understand each other.

Internet Image Search

☞ *Children's Games*
☞ *Hunters in the Snow*
☞ *The Peasant Dance*
☞ *The Fight between Carnival and Lent*
☞ *Procession to Calvary*
☞ *Two Chained Monkeys*
☞ *The Numbering at Bethlehem*

Student Instructions

1. Cut out the tiles from the jigsaw sheet.
2. Do the subtraction algorithms on the backing board.
3. Match the numbers on the tiles to the answers on the backing board.
4. Glue the tiles onto their matching spaces.
5. Color the picture in an interesting way.

Talking Mathematically

The students will need a good knowledge of regrouping to do these examples.

Our monetary system can be used to demonstrate "what's happening" when doing these subtractions; e.g., $196.05 – $134.20.

Follow these steps:

Step 1: Five one-cent coins minus 0 one-cent coins equals 5. Write 5 on the answer line under the one-cent column (the hundredths column).

Step 2: Look at the subtraction in the 10-cent column. Zero 10-cents minus two 10-cents can't be done—you must regroup (or borrow).

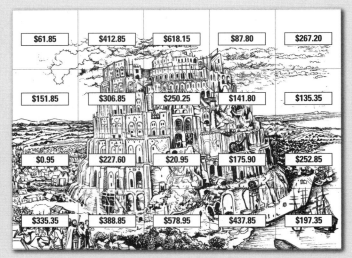

$61.85	$412.85	$618.15	$87.80	$267.20
$151.85	$306.85	$250.25	$141.80	$135.35
$0.95	$227.60	$20.95	$175.90	$252.85
$335.35	$388.85	$578.95	$437.85	$197.35

Step 3: Regroup or borrow from the dollar bills column. Exchange one of the 6 dollar bills for ten 10-cent coins. You now have 5 dollar bills in the dollar bills column (the units column) and ten 10-cents in the 10-cents column. Proceed with the 10-cents column subtraction—ten 10-cents minus two 10-cents equals eight 10-cents. Write 8 on the answer line under the ten-cent coins column.

Step 4: Do the subtraction in the dollar bills column. Five dollar bills minus 4 dollar bills equals 1 dollar bill. Write 1 on the answer line under the dollar bills column.

Step 5: Do the subtraction in the 10-dollar bills (tens) column. Nine 10-dollar bills minus three 10-dollar bills equals six 10-dollar bills. Write 6 on the answer line under the 10-dollar bills column.

Step 6: Finally, do the subtraction in the 100-dollar (hundreds) column. One 100-dollar bill minus one 100-dollar bill equals 0. There is no need to write 0 here as it does not add any meaning to the final answer.

Your students will need a good knowledge of regrouping to do these examples.

The Tower of Babel

$700.05 − 432.85	$378.85 − 243.50	$928.50 − 675.65	$402.05 − 204.70
$452.80 − 365.00	$819.95 − 678.15	$610.55 − 434.65	$616.65 − 178.80
$796.40 − 178.25	$928.05 − 677.80	$900.35 − 879.40	$769.80 − 190.85
$649.70 − 236.85	$674.10 − 367.25	$342.05 − 114.45	$677.70 − 288.85
$196.05 − 134.20	$919.65 − 767.80	$700.15 − 699.20	$812.95 − 477.60

MATH MASTERPIECES

The Tower of Babel

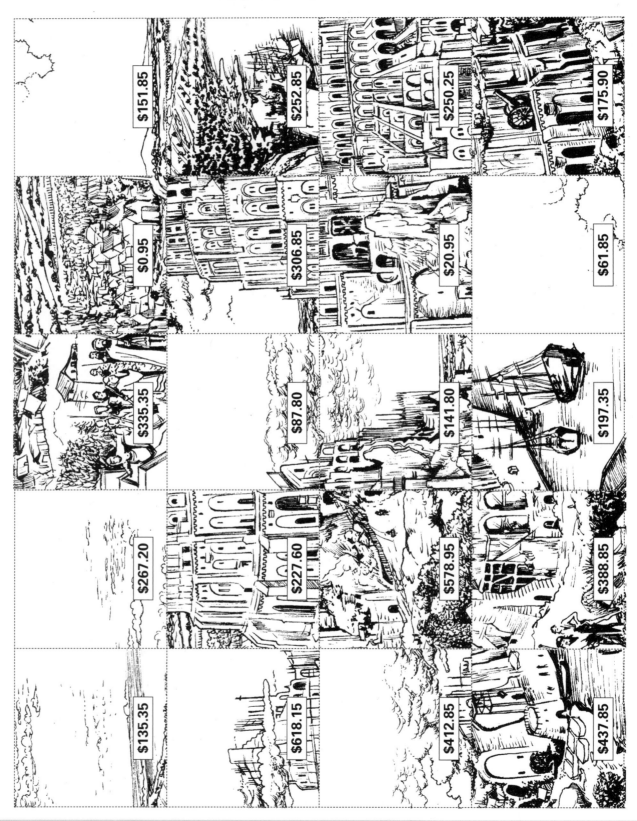

Max Schmitt in a Single Scull

Background Information

Max Schmitt in a Single Scull

by Thomas Eakins (1844–1916)

Thomas Eakins was an American painter who painted many scenes of outdoor sporting life. He was also a prolific painter of portraits. He often included himself as an observer in his paintings.

In *The Agnew Clinic* he is seen watching a surgical operation and in *The Swimming Hole* he is seen treading water next to his dog, Harry.

As an art teacher he stressed the importance of painting nude models and was ostracized as a result of this. Eakins was largely ignored by the art establishment during his lifetime, but is now highly regarded as one of America's greatest painters.

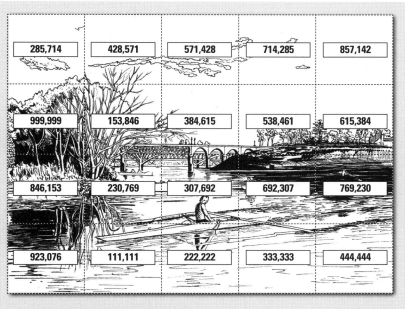

Internet Image Search

- ☞ *Starting Out after Rail*
- ☞ *Baseball Players Practicing*
- ☞ *The Agnew Clinic*
- ☞ *The Pathetic Song*
- ☞ *The Chess Players*

Student Instructions

1. Cut out the tiles from the jigsaw sheet.
2. Do the multiplication algorithms on the backing board.
3. Match the numbers on the tiles to the answers on the backing board.
4. Glue the tiles onto their matching spaces.
5. Color the picture in an interesting way.
6. Can you see why these numbers and their answers are amazing?

Talking Mathematically

These answers really are amazing.

Look at the answers when 142,857 is multiplied:

(x2) = 285,714   (x3) = 428,571   (x4) = 571,428

(x5) = 714,285   (x6) = 857,142

Not only does each answer contain the digits 142,857, they are also in order.

Such numbers are called roundabout numbers (the numbers 142,857 go around and around just as on a roundabout or, as they are better known, merry-go-round) and are extremely rare.

What is just as amazing is that a dedicated mathematician sat down and did hundreds of thousands of multiplication algorithms before coming across this oddity.

As if to show that it has more than one party trick, 142,857 x 7 makes 999,999.

76,923 when multiplied by 2, 5, 7, 8 and 11 gives the roundabout answers: 153,846, 384,615, 538,461, 615,384, 846,153.

When multiplied by 3, 4, 9, 10 and 12, a second set of roundabout answers results.

These are: 230,769, 307,692, 692,307, 769,230 and 923,076.

Max Schmitt in a Single Scull

142,857 × 6	76,923 × 8	76,923 × 10	63,492 × 7
142,857 × 5	76,923 × 7	76,923 × 9	47,619 × 7
142,857 × 4	76,923 × 5	76,923 × 4	31,746 × 7
142,857 × 3	76,923 × 2	76,923 × 3	15,873 × 7
142,857 × 2	142,857 × 7	76,923 × 11	76,923 × 12

Max Schmitt in a Single Scull

Study for Last Supper (Judas)

Background Information

Study for Last Supper (Judas)

by Leonardo da Vinci (1452–1519)

Leonardo da Vinci is regarded as the finest painter of the Renaissance (the period when art, learning and literature were revived [14th–16th centuries] after the period known as the Dark Ages). As well as being unsurpassed as a painter, Leonardo was also a highly accomplished engineer and scientist. Among his inventions are scissors.

This picture was a chalk drawing prepared as a model to be transferred for the famous painting, *The Last Supper*. Such line drawings were often done and transferred to canvas or a wall as a guide for the artist. Leonardo's master, the artist Verrochio, is said to have given up painting in awe when he saw the works of his 17-year-old student.

Perhaps your students can shade their finished masterpieces with rubbed chalk.

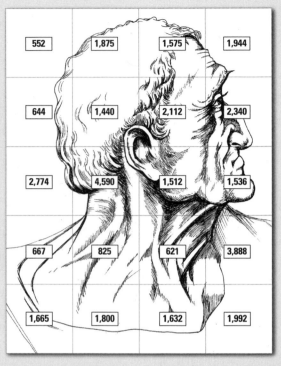

Internet Image Search

☞ *Mona Lisa*

☞ *The Last Supper*

☞ *Virgin of the Rocks*

Student Instructions

1. Cut out the tiles from the jigsaw sheet.

2. Do the multiplication algorithms on the backing board.

3. Match the numbers on the tiles to the answers on the backing board.

4. Glue the tiles onto their matching spaces.

5. Color the picture in an interesting way.

Talking mathematically

Work through a number of examples using these steps.

$$\begin{array}{r} 37 \\ \times\ 65 \\ \hline \\ \hline \\ \hline \end{array}$$

Step 1: Consider what "question" the algorithm is "asking." We are being asked how many 37 sets of 65 are altogether.

Step 2: Multiply 37 by 5 and write the answer (185) on the second line.

Step 3: Write 0 in the ones column on the third line because the next number you are multiplying by is 60 (not 6). Do the multiplication by 6 and write the answer on the third line (2220).

Step 4: Add the two answers (185 + 2220) together and write total on the bottom line (2405).

It will be useful if students have some experience multiplying by multiples of 10.

To "prove" the "write down zero" shortcut do some of these as additions. Children should accept the shortcut after doing a number of these; e.g., 20 x 40.

It should quickly become evident that the ones column adds up to 0. Just as clearly, students should quickly accept that 20 x 4 equals 80 (etc.) after working through a few examples using addition.

$$\begin{array}{r} 40 \\ \hline \\ \hline \end{array}$$

Study for Last Supper (Judas)

23 x 24	75 x 25	45 x 35	54 x 36
28 x 23	45 x 32	48 x 44	65 x 36
73 x 38	54 x 85	24 x 63	48 x 32
23 x 29	33 x 25	23 x 27	54 x 72
45 x 37	45 x 40	48 x 34	83 x 24

MATH MASTERPIECES

Study for Last Supper (Judas)

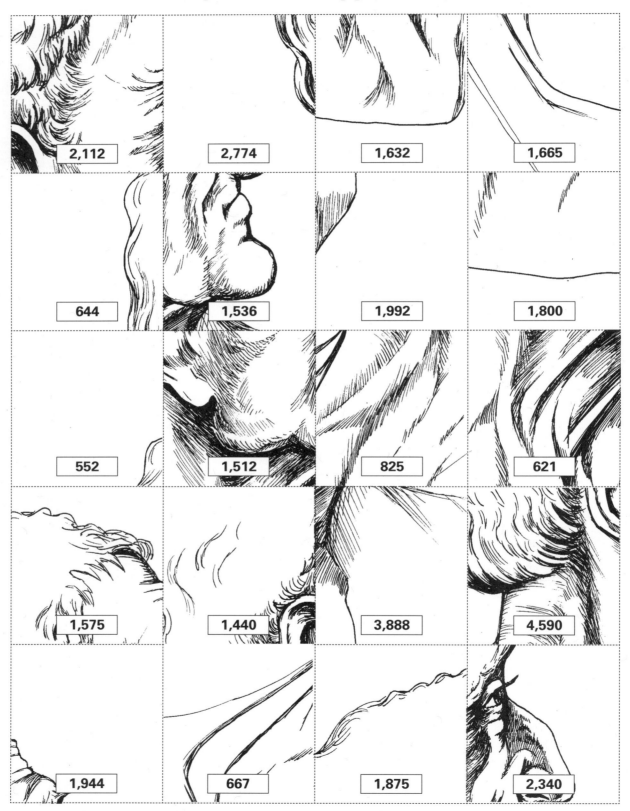

2,112 2,774 1,632 1,665

644 1,536 1,992 1,800

552 1,512 825 621

1,575 1,440 3,888 4,590

1,944 667 1,875 2,340

A Winter Scene

42 r 3	46 r 2	32 r 1	47 r 1
55 r 1	51 r 3	45 r 4	40 r 5
32 r 4	75 r 1	39 r 2	25 r 6
79 r 5	25 r 2	67 r 3	41 r 1
38 r 5	57 r 3	37 r 1	36 r 2

Background Information

A Winter Scene

by Isaac van Ostade (1621–1649)

Isaac van Ostade was a Dutch painter renowned for his winter scenes and paintings of peasant life in 17th century Holland.

His paintings are limited in number because he was only 28 years old when he died.

Internet Image Search

☞ *A Horse Stable*

☞ *A Peasant Pouring Beer in His Wife's Glass*

☞ *Peasants Outside a Farmhouse, Butchering Pork*

Student Instructions

1. Cut out the tiles from the jigsaw sheet.

2. Do the division algorithms on the backing board.

3. Match the numbers on the tiles to the answers on the backing board.

4. Glue the tiles onto their matching spaces.

5. Color the picture in an interesting way.

Talking mathematically

To assist with these algorithms initially, diagrammatic crutches may be provided to assist students with their calculations.

For example: $4\overline{)99}$

Diagram: tens: / / / / / / / / /

ones: / / / / / / / / /

Use the diagrams to demonstrate the first step, 9 divided by 4 (circle two groups of four tens). One will be one left uncircled. There are now 19 ones left (1 ten and 9 ones) which can be circled into four groups of 4, leaving three uncircled. This gives the answer 24 r 3.

MATH MASTERPIECES

A Winter Scene

$4\overline{)171}$	$3\overline{)140}$	$2\overline{)65}$	$2\overline{)95}$
$2\overline{)111}$	$4\overline{)207}$	$5\overline{)229}$	$8\overline{)325}$
$5\overline{)164}$	$2\overline{)151}$	$3\overline{)119}$	$8\overline{)206}$
$6\overline{)479}$	$3\overline{)77}$	$4\overline{)271}$	$2\overline{)83}$
$8\overline{)309}$	$5\overline{)288}$	$4\overline{)149}$	$3\overline{)110}$

A Winter Scene

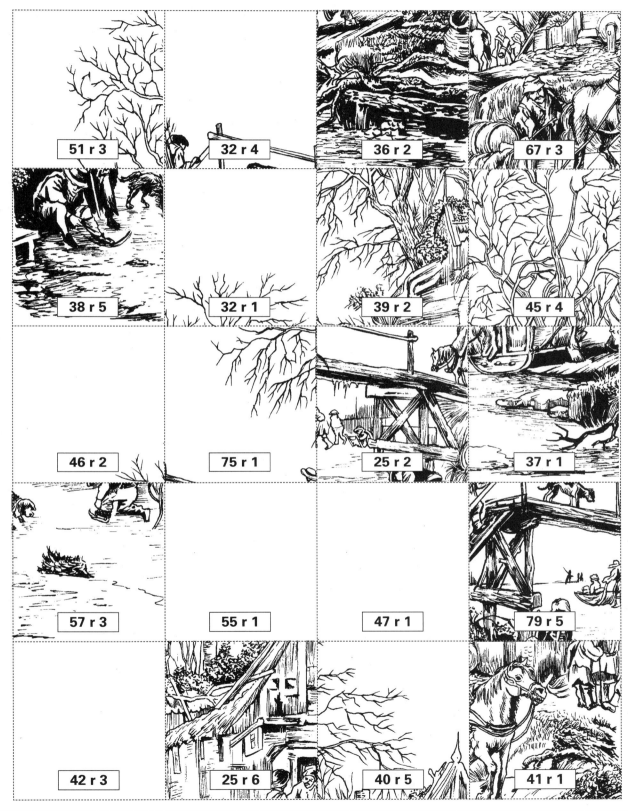

51 r 3   32 r 4   36 r 2   67 r 3

38 r 5   32 r 1   39 r 2   45 r 4

46 r 2   75 r 1   25 r 2   37 r 1

57 r 3   55 r 1   47 r 1   79 r 5

42 r 3   25 r 6   40 r 5   41 r 1

The Peasant Dance

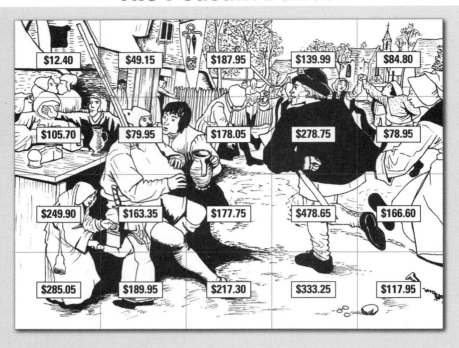

Background Information

The Peasant Dance

by Pieter Bruegel the Elder (1525–1569)

Pieter Bruegel the Elder is the best known of a family of Dutch artists. His paintings are very detailed and busy. *Children's Games* is a scene showing more than 200 children at play. Some people called him *Pieter the Peasant* because he liked to paint aspects of peasant life and would often mingle with peasants to get ideas for these paintings.

Internet image search

☞ *Children's Games*

☞ *Hunters in the Snow*

☞ *Peasant Wedding*

☞ *The Fight Between Carnival and Lent*

☞ *Procession to Calvary*

☞ *Two Chained Monkeys*

☞ *The Numbering at Bethlehem*

Student Instructions

1. Cut out the tiles from the jigsaw sheet.

2. Do the division algorithms on the backing board.

3. Match the amounts on the tiles to the answers on the backing board.

4. Glue the tiles onto their matching spaces.

5. Color the picture in an interesting way.

Talking Mathematically

Place the decimal point on the answer line above the decimal point of the dividend (number to be divided) before commencing the algorithm, and then proceed as with any other division algorithm.

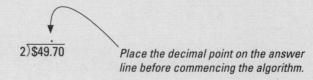

Place the decimal point on the answer line before commencing the algorithm.

The Peasant Dance

8)$678.40	5)$699.95	2)$375.90	4)$196.60	3)$37.20
9)$710.55	3)$836.25	5)$890.25	4)$319.80	6)$634.20
6)$999.60	2)$957.30	7)$1244.25	3)$490.05	4)$999.60
5)$589.75	3)$999.75	3)$651.90	4)$759.80	3)$855.15

MATH MASTERPIECES

The Peasant Dance

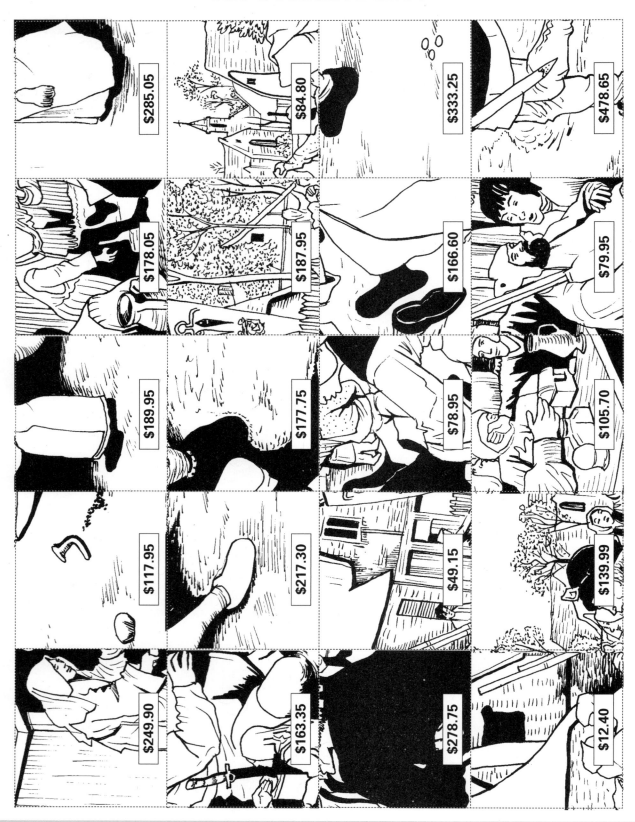

The Avenue at Middelharnis

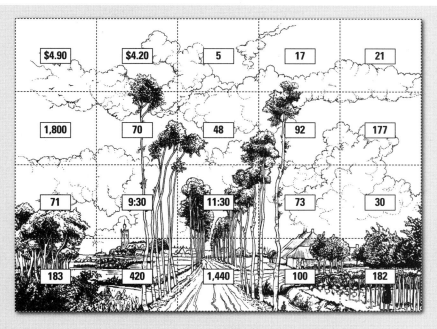

Background Information

The Avenue at Middelharnis

by Meindert Hobbema (1638–1709)

Meindert Hobbema was a Dutch artist who was not well known during his lifetime.

This picture shows the Dutch village of Middelharnis as it appeared around 1689. The painting is a good example of linear perspective.

Because he was unable to make a living from painting, Hobbema's output was not large. In 1688, he became a customs officer and was unable to devote as much time as he wished to painting.

Internet Image Search

☞ *The Watermill with a Red Roof*
☞ *A Stormy Landscape*

Student Instructions

1. Cut out the tiles from the jigsaw sheet.

2. Answer the problems on the backing board.

3. Match the numbers on the tiles to the answers on the backing board.

4. Glue the tiles onto their matching spaces.

5. Color the picture in an interesting way.

Talking Mathematically

Most of the real-life mathematics that the students will do will be in the form of mini-problems such as those listed here.

A useful framework for doing problem work is to ask the questions *What?* and *How?* and then *Answer*.

Example: 3 pens cost $2.10. How much will 7 cost?

What? What is the problem asking me to find out? (the cost of 7 pens.)

How? How do I work this out? What working will I need? (Find the cost of 1, then multiply this unit cost by 7.)

Answer $2.10 ÷ 3 = $0.70

7 x $0.70 = $4.90

Therefore, 7 pens will cost $4.90.

You may need to discuss some of the vocabulary used in these problems or provide resource materials for the students to find out some definitions. Some of the more difficult vocabulary used includes *score*, *gross* and *nonagon*.

The Avenue at Middelharnis

3 pens cost $2.10. How much for 7?	A dozen apples cost $7.20. How much for 7?	Plums cost $1.75 for 7. How many can you buy for $1.25?	Marbles sell at a price of 9 for 45¢. How many can Jocelyn buy for 85¢?	What is the perimeter of a rectangular room 7 yards long and 3.5 yards wide?
			——— yards
Tyler made 6 pies. Each pie had a mass of 300 ounces. What was the total mass he made? ——— ounces	How much is 3 score and 10?	How many legs altogether for 3 cows, 6 chickens, 3 fish and 4 ants?	How many days altogether in the months of winter?	How much is a gross plus a baker's dozen plus a score?
Two angles of a triangle are 47° and 62°. How big is the remaining angle? ——— °	Stanley has just finished work. He began 8 hours ago. It is 5:30 pm. When did he start? ——— am	A train trip took 5 and a quarter hours. It began at 6:15 am. When did it end? ——— am	How many sides do 4 hexagons, 8 pentagons and a nonagon have altogether?	Bruce's scores were 0, 1, 0, 0, 2 and 177. What was his average score?
How many days in autumn and spring combined?	A car travels 12 miles on a gallon of gasoline. How far will it go on 35 gallons? ——— miles	How many minutes in one day? ——— min	Water boils at this temperature. ——— °C	$3^2 + 4^2 + 6^2 + 11^2$ = ———

The Avenue at Middelharnis

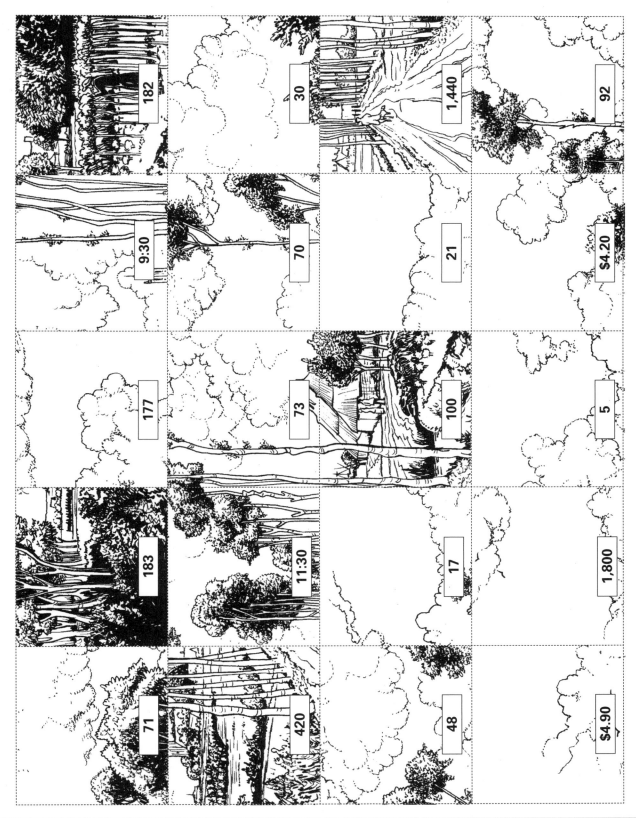

Centurion (from the Crucifixion of St. Peter)

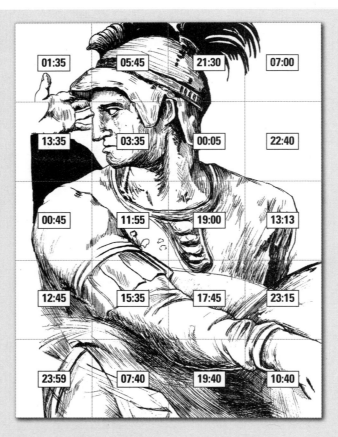

01:35	05:45	21:30	07:00
13:35	03:35	00:05	22:40
00:45	11:55	19:00	13:13
12:45	15:35	17:45	23:15
23:59	07:40	19:40	10:40

Background Information

Centurion (from the Crucifixion of St Peter)

by Michelangelo Buonarroti (1475–1564)

Michelangelo is regarded as one of the greatest artists of all time and was equally gifted as a painter and a sculptor.

Among his many great works is the ceiling of the Vatican's Sistine Chapel. This took him about four years and depicts many scenes from the Bible, including three stories about Adam and Eve and three about Noah.

Internet Image Search

☞ *Sistine Chapel Ceiling (many depictions of nudity)*

Student Instructions

1. Cut out the tiles from the jigsaw sheet.

2. Match the 24-hour times on the tiles to their equivalent 12-hour times on the backing board.

3. Glue the tiles onto their matching spaces.

4. Color the picture in an interesting way.

Talking Mathematically

Why do we divide a day into two divisions of 12 hours?

It is believed that clock makers were influential in causing this to come about. Old clocks worked on a spring mechanism. When the spring runs down it has to be rewound. A 12-hour clock chimes 156 times in 24 hours (two lots of 78 chimes). A clock chiming up to 24 times a day would chime a total of 300 times in a day and so use more power from its spring. Such a clock would have to be rewound more frequently.

Clock faces with 24 divisions on the clock face are hard to read with a numeral being placed every 15°. On a 12-hour clock face a numeral is only necessary every 30°.

However, 24-hour time is still popular and used in scheduling by, for example, railways and airlines.

Centurion (from the Crucifixion of St. Peter)

1:35 a.m.	a quarter to six in the morning	half past nine at night	7:00 a.m.
1:35 p.m.	twenty-five minutes to four in the morning	five minutes past midnight	twenty to eleven at night
12:45 a.m.	five minutes before noon	7:00 p.m.	thirteen minutes past one in the afternoon
12:45 p.m.	twenty-five minutes to four in the afternoon	a quarter to six in the afternoon	three quarters of an hour before midnight
one minute to midnight	twenty minutes to eight in the morning	twenty minutes to eight in the evening	twenty to eleven in the morning

Centurion (from the Crucifixion of St. Peter)

The Miraculous Draught of Fishes

Background Information

The Miraculous Draught of Fishes

by Konrad Witz (c.1400–1444)

Living at different times in Germany and Switzerland, little is known of Witz's life. He was one of the first artists to incorporate realistic landscapes into religious paintings.

Internet Image Search

☞ *Saint Bartholomew*

☞ *The Queen of Sheba before Solomon*

☞ *Saint Christophorus*

☞ *Madonna and Saints in a Church*

☞ *Meeting of Joachim and Anna by the Golden Gate*

Student Instructions

1. Cut out the tiles from the jigsaw sheet.

2. Do the multiplication algorithms on the backing board.

3. Match the numbers on the tiles to the fractions on the backing board.

4. Glue the tiles onto their matching spaces.

5. Color the picture in an interesting way.

Talking Mathematically

It is important that students know the meaning of such algorithms before the "shortcuts" are taught and applied. For example, $3/8 \times 64$, should be read as "three-eighths of sixty-four" (not three-eighths times sixty-four) and explained as meaning "three out of every eight in a total group of sixty-four."

The mechanical process involved in finding the answer is to circle groups of eight and then mark (color, cross, highlight) three out of every such group (i.e., three out of every eight in a total group of 64). Finally, when each group of eight has been thus marked, count the total number of items marked.

After doing this a number of times, some students will see that a "shortcut" exists; i.e., divide the denominator of the fraction into the total amount, then multiply your answer by the numerator.

MATH MASTERPIECES

The Miraculous Draught of Fishes

$\frac{2}{3} \times 27 =$

$36 \times \frac{3}{4} =$

$\frac{11}{12} \times 36 =$

$49 \times \frac{3}{7} =$

$\frac{2}{7} \times 28 =$

$\frac{7}{8} \times 48 =$

$\frac{8}{11} \times 44 =$

$\frac{5}{7} \times 56 =$

$\frac{2}{3} \times 30 =$

$\frac{3}{5} \times 20 =$

$\frac{4}{7} \times 63 =$

$\frac{2}{5} \times 25 =$

$36 \times \frac{4}{9} =$

$40 \times \frac{5}{8} =$

$42 \times \frac{5}{6} =$

$\frac{1}{7} \times 49 =$

$\frac{3}{8} \times 64 =$

$\frac{1}{4} \times 24 =$

$\frac{4}{5} \times 35 =$

$\frac{3}{7} \times 21 =$

The Miraculous Draught of Fishes

The Feast of Belshazzar

Background Information

The Feast of Belshazzar

by Rembrandt van Rijn (1606–1669)

The Dutch painter Rembrandt van Rijn is regarded as one of the greatest artists of all time. Although highly regarded during his lifetime, he never attained great wealth and was declared bankrupt in 1656.

As well as portrait painting he chose religious themes as subjects for his paintings. *The Feast of Belshazzar* recounts a scene from the Bible story of Belshazzar, the King of Babylon (555 to 539 BC). It is said that during a feast held by Belshazzar a hand was seen to write on the wall foretelling the destruction of Babylon. Babylon did fall a short time later and Belshazzar was killed.

Internet Image Search

☞ *Woman Bathing in a Stream*
☞ *The Anatomy Lesson of Dr. Tulp*
☞ *Tobit and Anna*
☞ *The Ass of Balaam Talking Before the Angel*
☞ *Christ in the Storm on the Lake Galilee*
☞ *Saskia*
☞ *Self-Portrait*
   (Rembrandt did many of these at various times of his life.)

Student Instructions

1. Cut out the tiles from the jigsaw sheet.
2. Draw lines to form equivalent fractions on the backing board.
3. Match the fractions on the tiles with the answers on the backing board.
4. Glue the tiles onto their matching spaces.
5. Color the picture in an interesting way.

Talking Mathematically

Just as it is legitimate to write the whole number 7 as 1 + 1 + 1 + 4, (3 x 2) + 1 or in any number of equivalent ways, so too can fractions be represented in any number of ways using what are called equivalent fractions.

Students are usually taught the shortcut method of forming equivalent fractions (multiply numerator and denominator by the same amount) without understanding what this means.

A simple way of explaining "what happens" is to draw shapes (cakes) as shown in the diagrams and "cut them up differently."

For example:

Ask students "What fraction of the 'cake' above has dark icing?" ($^2/_5$)

Place dots at points A, B, C and D and ask students to draw "cut lines" from A to B and from C to D.

Ask "What fraction has dark icing now that the cake has been cut differently?" ($^6/_{15}$)

Ask "Has any icing been added to the cake?" (No)

Ask "Has any icing been removed?" (No)

Ask "If no icing has been added and none removed, are $^2/_5$ and $^6/_{15}$ the same amount?" (Yes)

Explain: Yes, they are the same amount, but the cakes have just been cut up differently. That is a useful way to think of equivalent fractions.

Why does the shortcut method work?

When you multiply the numerator and denominator by 3 you are effectively saying that three-thirds (one whole cut into thirds) of two-fifths is six-fifteenths. You are describing "one whole lot of fifths" but the one whole is shown as three-thirds.

The Feast of Belshazzar

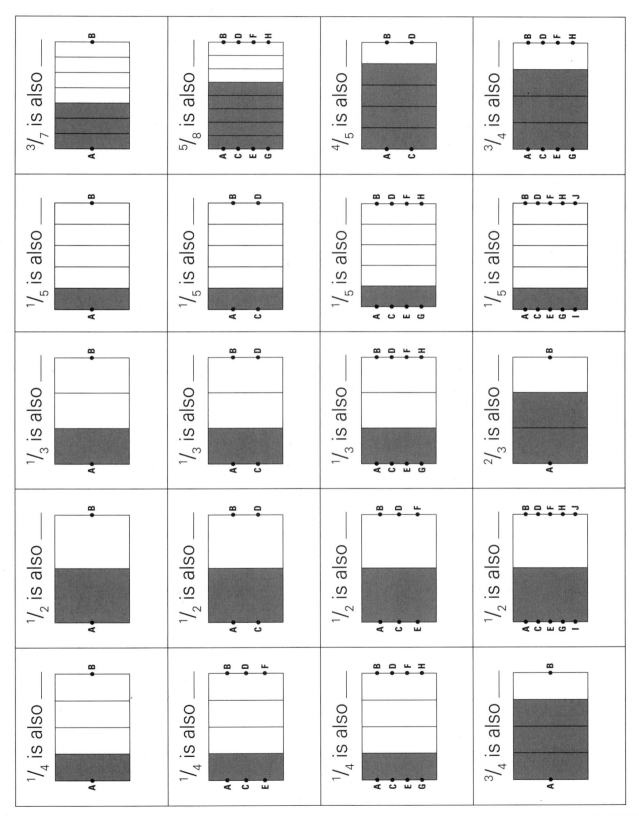

MATH MASTERPIECES

The Feast of Belshazzar

Bailed Up

Background Information

Bailed Up

by Tom Roberts (1856–1931)

Tom Roberts was an important Australian artist of the latter half of the 19th and early 20th centuries. Roberts thought it was important to record Australia's pioneering days.

Bailed Up depicts a true incident described to Roberts by a friend, Bob Bates. Bates had been the stagecoach driver of a Cobb & Co. coach traveling between Glen Innes and Inverell when it was held up. Roberts created the painting from these recollections.

Such hold ups were all too common in early Australia. As well as the human drama, Roberts shows an equally dramatic hot dry landscape, typical of Australia in midsummer.

Internet Image Search

☞ *Bourke Street (Allegro Con Brio)*

☞ *Coming South*

☞ *Shearing the Rams*

☞ *A Break Away!*

Student instructions

1. Cut out the tiles from the jigsaw sheet.

2. Change the fractions on the backing board to their decimal equivalents.

3. Match the numbers on the tiles to the answers on the backing board.

4. Glue the tiles onto their matching spaces.

5. Color the picture in an interesting way.

Talking Mathematically

The students will require a knowledge of equivalent fractions to do this activity. Most of the fractions chosen can be easily converted to tenths or hundredths by multiplying numerators and denominators by equal amounts.

Some fractions, however, do not convert so easily to tenths, hundredths, thousandths, etc.

Here is a method you can use to convert these more difficult examples to decimals:

$$^1/_3$$

Step 1: Divide the top number by the bottom number.

(Three goes into one zero times and one is "left over.")

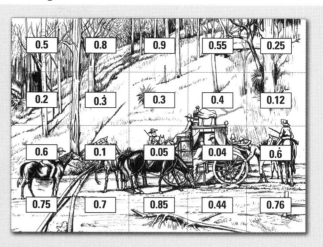

Step 2: Write down the zero and put a decimal point after it.

(In other words one third is less than one whole.)

Step 3: Add zero to the one left over and divide again by the bottom number.

(Three goes in to ten three times with one left over.)

Write the three after the decimal point.

0.3

From now on, whenever there is a remainder you must add zero to it and divide again by the bottom number.

Working will look something like this:

$$^{1\,0}/_3 = 10 \quad 10$$

zero *"written in"*   left over   left over

zeroes *"written in"*

0.33

As you can see, the answer we will continue to get is 3. The fraction $^1/_3$ forms the decimal *zero point three recurring*.

No matter how long you continue to do the division, you will continue to get the answer 3.

To show this, it is written $0.\dot{3}$ in decimal notation.

For the fraction $^1/_7$, 142,857 continues to repeat. It is called zero point one four two eight five seven recurring and written $0.\dot{1}4285\dot{7}$ in decimal notation.

For the fraction $^1/_6$ the 6 repeats. We write this $0.1\dot{6}$ in decimal notation.

Bailed Up

$\frac{1}{4} = 0.$ ___	$\frac{3}{25} = 0.$ ___	$\frac{2}{3} = 0.$ ___	$\frac{19}{25} = 0.$ ___
$\frac{11}{20} = 0.$ ___	$\frac{2}{5} = 0.$ ___	$\frac{1}{25} = 0.$ ___	$\frac{11}{25} = 0.$ ___
$\frac{9}{10} = 0.$ ___	$\frac{3}{10} = 0.$ ___	$\frac{1}{20} = 0.$ ___	$\frac{17}{20} = 0.$ ___
$\frac{4}{5} = 0.$ ___	$\frac{1}{3} = 0.$ ___	$\frac{1}{10} = 0.$ ___	$\frac{7}{10} = 0.$ ___
$\frac{1}{2} = 0.$ ___	$\frac{1}{5} = 0.$ ___	$\frac{3}{5} = 0.$ ___	$\frac{3}{4} = 0.$ ___

Bailed Up

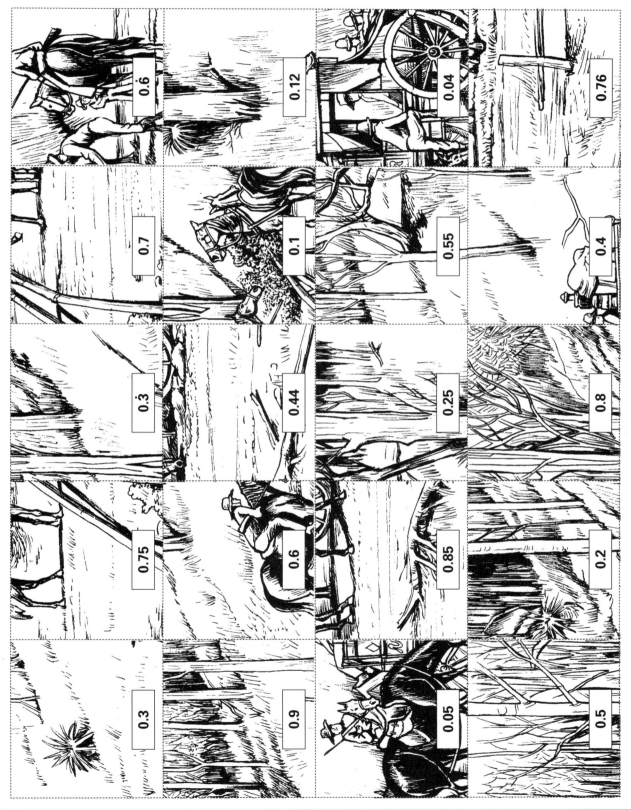

Bathers at Asnières

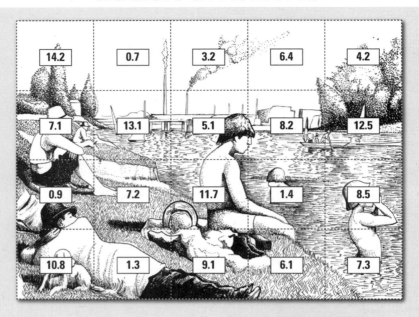

14.2	0.7	3.2	6.4	4.2
7.1	13.1	5.1	8.2	12.5
0.9	7.2	11.7	1.4	8.5
10.8	1.3	9.1	6.1	7.3

Background Information

Bathers at Asnières

by Georges Seurat (1859–1891)

Georges Seurat was a French artist who developed a method of painting called *pointillism* (also *divisionism*). This involved creating his paintings with a series of tiny dots so the colors are mixed by the eye (tiny red and yellow dots placed close together on the canvas would be mixed by the eye to make the color orange, etc.). Color television produces colored images in much the same way, colors being mixed by the eyes using tiny red, green and blue dots.

Internet Image Search

☞ *Sunday Afternoon on the Island of Grande Jatte*
☞ *The Can-Can*
☞ *Rock-Breakers*
☞ *Cows in a Field*
☞ *Watering Can*
☞ *Invitation to the Sideshow*
☞ *The Eiffel Tower*

Student Instructions

1. Cut out the tiles from the jigsaw sheet.

2. Do the addition algorithms on the backing board.

3. Match the numbers on the tiles to the answers on the backing board.

4. Glue the tiles onto their matching spaces.

5. Color the picture in an interesting way.

Talking Mathematically

It is important to be aware of place value when dealing with decimals, just as it is with whole numbers.

Bathers at Asnières

2.8 + 1.4	7.7 + 4.8	6.9 + 1.6	4.7 + 2.6
3.3 + 3.1	2.6 + 5.6	0.9 + 0.5	3.5 + 2.6
1.9 + 1.3	3.9 + 1.2	5.6 + 6.1	1.2 + 7.9
0.2 + 0.5	7.8 + 5.3	3.7 + 3.5	0.5 + 0.8
5.9 + 8.3	3.9 + 3.2	0.1 + 0.8	3.9 + 6.9

MATH MASTERPIECES

Bathers at Asnières

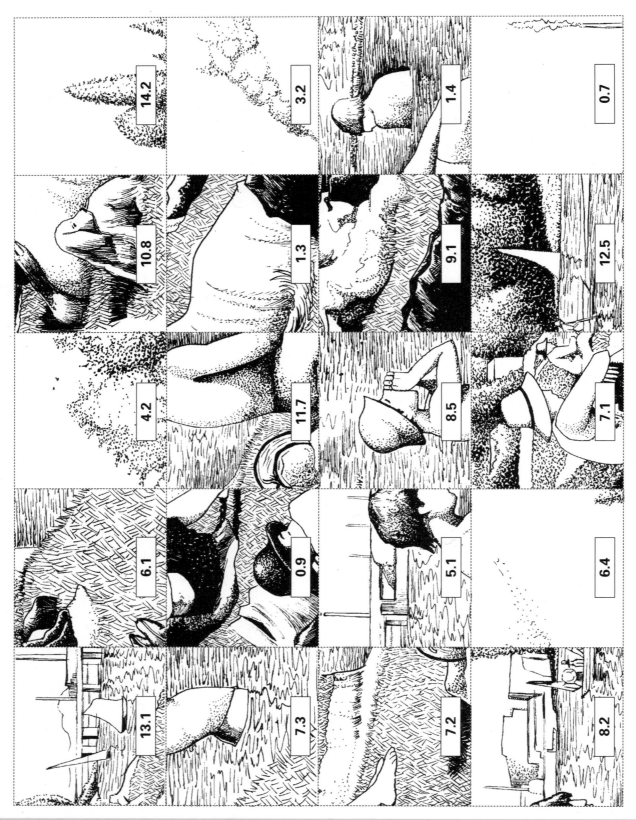

The Scapegoat

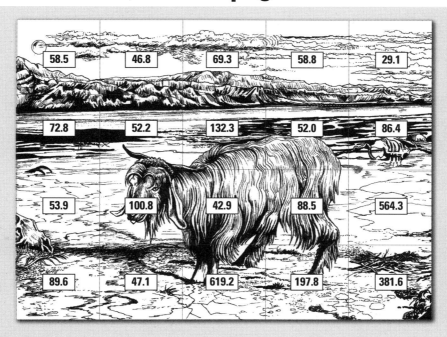

Background information

The Scapegoat

by William Holman Hunt (1827–1910)

William Holman Hunt was an Englishman who studied art at Britain's famous school of art, The Royal Academy. During his period there he became friends with the artists John Everett Millais and Dante Gabriel Rossetti.

The three artists founded a group and a painting style known as the *Pre-Raphaelite Brotherhood*. Their aim was to have serious moral issues as the subject of their paintings and to represent nature with realism.

Hunt went to Palestine to work on a series of religious paintings in 1854. It was here that he painted the nightmarish scene *The Scapegoat*. It depicts the Bible story of the goat cast into the wilderness by a Jewish chief priest who had symbolically laid people's sins upon it.

The word *scapegoat* has come to mean a person unfairly blamed for the wrongdoings of others, just as the goat was in the Bible story.

Internet Image Search

☞ *The Light of the World*
☞ *The Hireling Shepherd*

Student Instructions

1. Cut out the tiles from the jigsaw sheet.

2. Do the multiplication algorithms on the backing board.

3. Match the numbers on the tiles to the answers on the backing board.

4. Glue the tiles onto their matching spaces.

5. Color the picture in an interesting way.

Talking Mathematically

It is important to be aware of place value when doing these multiplications. The decimal point can be placed between the answer lines before any work is done. Then proceed with the multiplication as if the decimal point was not present.

The students can check the placement of the decimal point in the answer by rounding any decimal numbers in the question to whole numbers and mentally multiplying them. For example, 6.5 x 9 ≈ 7 x 9; therefore, the answer should be close to 63.

The Scapegoat

9.7 × 3	8.4 × 7	9.9 × 7	7.8 × 6	6.5 × 9
9.6 × 9	6.5 × 8	18.9 × 7	17.4 × 3	9.1 × 8
62.7 × 9	17.7 × 5	14.3 × 3	12.6 × 8	7.7 × 7
63.6 × 6	98.9 × 2	77.4 × 8	15.7 × 3	11.2 × 8

The Scapegoat

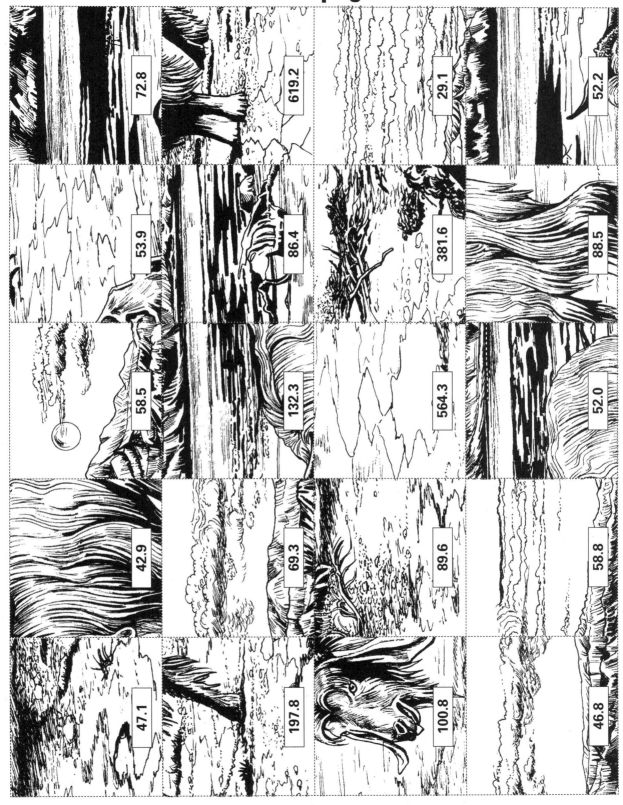

Self-Portrait

Background Information

Self-Portrait

by Vincent van Gogh (1853–1890)

Vincent van Gogh was a brilliant artist but led an unsettled life because of recurring periods of mental disturbance. He only began painting at the age of 27, but in the ten-year period of activity before he shot and killed himself he completed 1,600 paintings.

He was not recognized until after his death, and most of his life was lived in abject poverty. He was a friend of the French artist, Paul Gauguin, and it was after an argument with him that van Gogh cut off part of his own ear.

Most of his paintings are made up of separate unblended brush marks. Van Gogh always signed his pictures with his first name, Vincent, only.

This is one of several self-portraits painted by the artist in which we see him as an intense, ravaged, restless-looking young man.

Internet Image Search

☞ *Sunflowers*
☞ *Seashore at Scheveningen*
☞ *The State Lottery Office*
☞ *The Potato Eaters*
☞ *Agostina Segatori*
☞ *The Good Samaritan*
☞ *The All Night Café at Arles*
☞ *The Dance Hall at Arles*

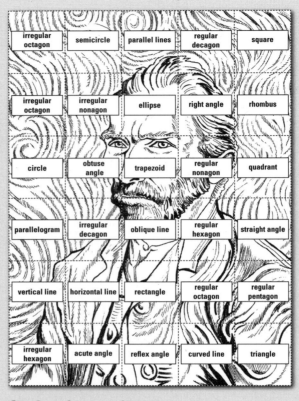

Student Instructions

1. Cut out the tiles from the jigsaw sheet.

2. Match the words on the tiles to their pictures on the backing board.

3. Glue the tiles onto their matching spaces.

4. Color the picture in an interesting way.

Talking Mathematically

The students will need to understand the difference between regular and irregular shapes to complete this activity.

Self-Portrait

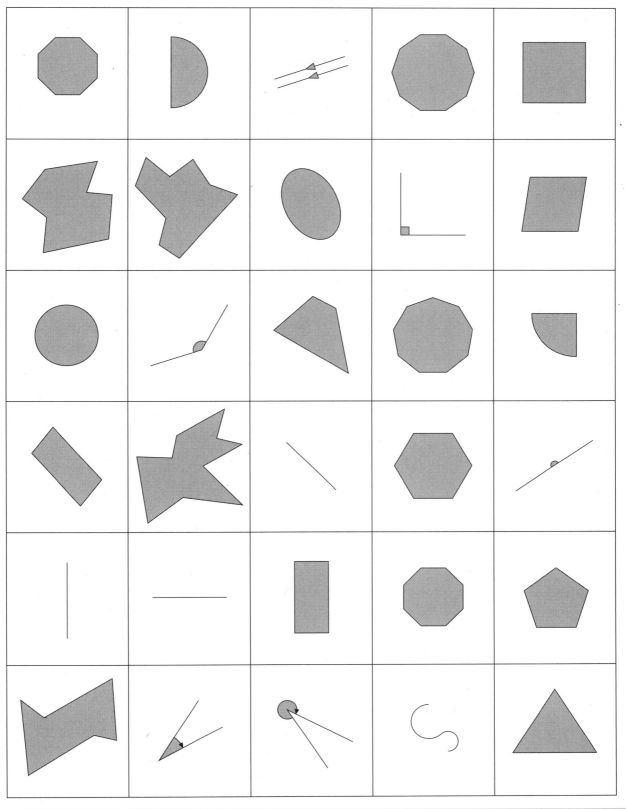

MATH MASTERPIECES

Self-Portrait

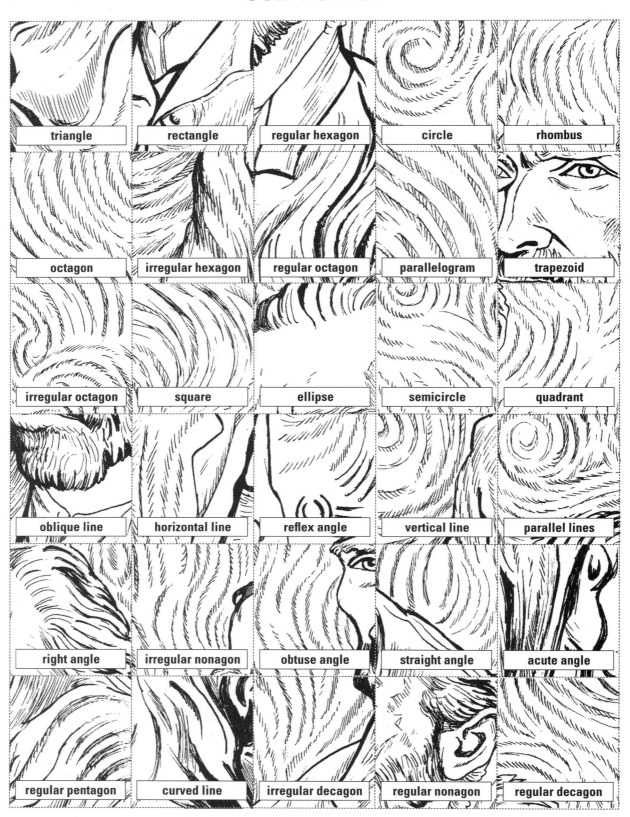

triangle	rectangle	regular hexagon	circle	rhombus
octagon	irregular hexagon	regular octagon	parallelogram	trapezoid
irregular octagon	square	ellipse	semicircle	quadrant
oblique line	horizontal line	reflex angle	vertical line	parallel lines
right angle	irregular nonagon	obtuse angle	straight angle	acute angle
regular pentagon	curved line	irregular decagon	regular nonagon	regular decagon

Number intuition

Multiplying odd and even numbers

Teacher information

We often ask children to consider the following patterns.

- *odd x odd = odd*
- *odd x even = even**
- *even x even = even*

* *Multiplying an odd number by an even number will produce the same result as multiplying an even number by an odd number. Students should explore both situations and discover this for themselves.*

Rarely do children apply what they learn about these patterns as a means of checking a calculation.

These checks can be stated more simply.

> - *The answer to a multiplication can only be odd if two odd numbers are multiplied.*

How will learning about odd and even numbers help in checking a calculation? It is easy to tell whether a number is odd or even simply by looking at the ones digit. Once you are aware of the pattern, all you need to do is inspect the ones digits of the numbers that are multiplied, and the answer, to determine whether the answer is definitely wrong or possibly correct.

Answers

1. (a) *12*
   (b) *96*
   (c) *1768*
   (d) *Teacher check*
   (e) *Teacher check*

2. *When two even numbers are multiplied an even result is produced.*
   even x even = even

3. (a) *33*
   (b) *351*
   (c) *7275*
   (d) *Teacher check*
   (e) *Teacher check*

4. *When two odd numbers are multiplied an odd result is produced.*
   odd x odd = odd

5. (a) *52*
   (b) *108*
   (c) *962*
   (d) *Teacher check*
   (e) *Teacher check*

6. *When an even and an odd number are multiplied an even result is produced.*
   even x odd = even

7. *Teacher check*
   even x even x even = even
   odd x odd x odd = odd

   When you multiply three numbers, all three must be odd in order for the answer to be odd.

   The order in which the multiplication occurs does not matter. The result is the same.

Number intuition # Multiplying odd and even numbers

1. Investigate what happens when two even numbers are multiplied.

    (a) 2 x 6 = ____

    (b) 12 x 8 = ____

    (c) 34 x 52 = ____

    How will learning about odd and even numbers help in checking the answer to a calculation?

    (d) ____ x ____ = ____

    (e) ____ x ____ = ____

2. What do you notice about the answer when two even numbers are multiplied?

    _____

    _____

    _____

3. Investigate what happens when two odd numbers are multiplied.

    (a) 3 x 11 = ____

    (b) 13 x 27 = ____

    (c) 75 x 97 = ____

    (d) ____ x ____ = ____

    (e) ____ x ____ = ____

4. What do you notice about the answer when an even and odd number are multiplied?

    _____

    _____

    _____

5. Investigate what happens if an odd and even number are multiplied.

    (a) 4 x 13 = ____   (d) ____ x ____ = ____

    (b) 12 x 9 = ____   (e) ____ x ____ = ____

    (c) 26 x 37 = ____

6. What do you notice about the answer when an even and odd number are multiplied?

    _____

    _____

    _____

    _____

    _____

7. Will the rules you discovered for multiplying odd and even numbers still work if more than two numbers are multiplied? Write questions to test various possibilities. For example, you might like to investigate:

    • *What happens when you multiply three even numbers or three odd numbers?*

    • *What happens when odd and even numbers are mixed, for example, odd x even x even?*

    • *Does the order in which odd and even numbers are multiplied make a difference; for example, even x even x odd or even x odd x even?*

    Explain your findings on the back of this sheet.